# Health Communications in Traditional Chinese Medicine Cardiology

# Health Communications in Traditional Chinese Medicine Cardiology

Dr. Anika Niambi Al-Shura, BSc., MSOM, Ph.D
Continuing Education Instructor
Niambi Wellness
Tampa, FL

Medical Illustrator: Samar Sobhy

AMSTERDAM • BOSTON • HEIDELBERG • LONDON
NEW YORK • OXFORD • PARIS • SAN DIEGO
SAN FRANCISCO • SINGAPORE • SYDNEY • TOKYO

ELSEVIER

Academic Press is an imprint of Elsevier

Academic Press is an imprint of Elsevier
32 Jamestown Road, London NW1 7BY, UK
The Boulevard, Langford Lane, Kidlington, Oxford, OX5 1GB, UK
Radarweg 29, PO Box 211, 1000 AE Amsterdam, The Netherlands
225 Wyman Street, Waltham, MA 02451, USA
525 B Street, Suite 1900, San Diego, CA 92101-4495, USA

**Notices**
Knowledge and best practice in this field are constantly changing. As new research and
experience broaden our understanding, changes in research methods, professional practices, or
medical treatment may become necessary.

Practitioners and researchers must always rely on their own experience and knowledge in
evaluating and using any information, methods, compounds, or experiments described herein.
In using such information or methods they should be mindful of their own safety and the safety
of others, including parties for whom they have a professional responsibility.

To the fullest extent of the law, neither the Publisher nor the authors, contributors, or editors,
assume any liability for any injury and/or damage to persons or property as a matter of products
liability, negligence or otherwise, or from any use or operation of any methods, products,
instructions, or ideas contained in the material herein.

**British Library Cataloguing in Publication Data**
A catalogue record for this book is available from the British Library

**Library of Congress Cataloging-in-Publication Data**
A catalog record for this book is available from the Library of Congress

ISBN: 978-0-12-800125-7

For information on all Academic Press publications
visit our website at **store.elsevier.com**

This book has been manufactured using Print On Demand technology. Each copy is produced to order
and is limited to black ink. The online version of this book will show color figures where appropriate.

# DEDICATION

The energy and effort behind the research and writing of this textbook is dedicated to my son, Khaleel Shakeer Ryland. May this inspire and guide you through your journey in your medical studies, career, and life.

# ACKNOWLEDGMENTS

This is a special acknowledgment to my seven-year medical students at Tianjin Medical University (2012–2013) who served as cardiovascular research assistants. May your future medical careers be successful.

An Qi He
Bin Lin Da
Han Jiang
Chen Hua
Jia Ying Luo
Jun Zhang
Lin Lin
Ming Lu
Nang Zhang
Ping Tang
Hu Si Le
Zhao Tian Man
Wen Xing Ning
Xing Wen Zhao
Tang Ying Mei
Li Ying Ying
Xiong Yong Qin
Ding Yu
Li Yan Jun

# CONTENTS

# SECTION III  CASELOAD AND RESEARCH

The companion course which is required for study with this textbook edition can be found on the Elsevier website and at www.niambiwellness.com.

## APPROVING AGENCIES

The course #1053-001 is approved by the National Certification Commission for Acupuncture and Oriental Medicine (NCCAOM) for 9 PDA points.

## COURSE DESCRIPTION

This course provides an overview of the history of ancient and modern public health, the importance of Oriental medicine, and role of the integrative medicine health provider.

## COURSE OBJECTIVES

- Analyze the history, key elements, and scientific theories to realize their importance in integrative Oriental medicine practice.
- Identify communication steps to enhance practitioner and patient relationships.
- Explain simple research methods to improve over-all patient care.

# SECTION *I*

# Rationale for Health Communications

# CHAPTER 1

# History of Health Communications

## CHAPTER OBJECTIVES

After studying this chapter, you should be able to:

1. Explain the overview of notable periods in world history when Chinese medicine and Western medicine responded to emergencies in public health, which we still use today.
2. Name the scientists and their medical discoveries.
3. Describe the basics of health communication and the significance of health literacy.

## 1.1 PART 1: HISTORY OF PUBLIC HEALTH

### 1.1.1 Lesson 1: History of Public Health

The perspective of public health practices in the Western world is influenced more by recent discoveries within the past 300 years. The following is a brief overview of a few periods, events, scientists, medicines, and devices which are part of our modern ways of living.

#### 1.1.1.1 Notables of Ancient China

1.1.1.1.1 Huang di Nei jing

The Huang di Nei jing (Nei jing) is one of the oldest major medical texts. It is said to have been written between the late Warring States period (475–221 BCE) and the early Han period (206–220 BCE). It is considered the most influential Chinese medicine text, which first established the ancient science and art of Traditional Chinese Medicine. It includes two major texts:

1. Su Wen/the Nei jing (Plain Questions)
2. The Ling Shu (Spiritual Pivot)

Many sections of the original books are now missing. However, the Nei jing is thought to have been modified many times by different scholars over many centuries. The Su Wen first described the aspects

Health Communications in Traditional Chinese Medicine Cardiology. DOI: http://dx.doi.org/10.1016/B978-0-12-800125-7.00001-X

of disease prevention. According to the Su Wen, a special formula was used to keep pestilence away.

1.1.1.1.2 Hua Tuo

Hua Tuo was an ancient physician, surgeon, and anesthesiologist. He lived during the late Han Dynasty (206 BC–220 AD) and Three Kingdoms era (220–280 AD) of Chinese history. He was considered to be a conscientious physician, who was concerned with serving his patients well. One of his main concerns in medicine was for public health and disease prevention. Hua Tuo created many decoctions which used only a few herbs. He was well known for very interesting decoctions.

Hua Tuo created a certain medical law. The Dao Yin medical exercises that are also called the five-mimic animal exercises follow this law. All of the texts he wrote which describe his techniques, prescriptions, and protocols are now considered lost.

---

**Section Assessment**

1. When was the Nei jing written?
2. What are the two books which are part of it?
3. According to the companion course, what did the Su Wen describe?
4. According to the companion course, what were the medicines created by Hua Tuo for?
5. According to the companion course, what is the name of the medical law created by Hua Tuo?

---

## 1.1.1.2 Pandemics and Epidemics

1.1.1.2.1 Spontaneous Generation

This theory is associated with teachings of Anaximander 610–546 BC.

1.1.1.2.2 Miasma Theory

This theory was the popular explanation for pandemics for many centuries. It suggested an interesting origin of epidemic diseases. The belief was that they came from the vapors of purulent matter which spread through air and water.

The smallpox pandemic was first thought to come from a miasma. Smallpox as a pandemic occurred during different periods in ancient

world history. At least a couple of diseases are thought to have origi-
nated from China over the past 1000 years. The early assumption was
that miasma from China affected traders along the Silk Road, and first
brought smallpox into Europe. Two common herbs used today were
instrumental in the disinfections.

The miasmic theory eventually influenced public health practices in
Europe, the Middle East and Asia. The plagues in Europe from the
1300s to early 1900s resulted in the death of millions of people. During
early periods, the dead were left where they lay and towns were aban-
doned. Miasma of the infection was thought to accumulate and spread
as a result.

Later, it was believed that somehow containing the disease in a
single isolated location was a good solution. Mass graves were
organized to prevent the spread of infection. The development of a
quarantine system and burning entire towns were also thought to stop
the spread. A significant discovery happened during a plague period in
China which changed medical practices forever.

## 1.1.1.2.3 Germ Theory
This theory states that microorganisms can invade living hosts and
cause a disease.

In Europe, the miasma theory was replaced by the germ theory of
disease during the late 1800s with discoveries and developments by
Pasteur, Agostino Bassi, Ignaz Semmelweis, and Robert Koch. Robert
Koch's postulates have influenced modern developments in pathogene-
sis to help guide gene encoding.

| Koch's postulates |
| --- |
| 1. The microorganism must be found in abundance in all organisms suffering from the disease, but should not be found in healthy organisms. |
| 2. The microorganism must be isolated from a diseased organism and grown in pure culture. |
| 3. The cultured microorganism should cause disease when introduced into a healthy organism. |
| 4. The microorganism must be re-isolated from the inoculated. |

## 1.1.1.2.4 Medical Scientists and Their Discoveries

| Scientist | Year | 19th Century Developments |
|---|---|---|
| John Snow | 1854 | Snow is famous for investigating the disease source which ended the cholera plague in London. He further developed anesthesia, hygiene and epidemiology in Europe. |
| Robert Koch | 1880's | Koch developed postulates to match organisms to diseases. His research included anthrax, cholera and tuberculosis. |
| Louis Pasteur | 1880's | Pasteur's discoveries involving vaccinations began with the development of serums by fermentation and pasteurization. |
| Semmelweis and Lister | 1870's | Both helped to develop antiseptics for the clinical and hospital setting. |
| Alexander Flemming | 1928 | Flemming developed antibiotics, yet they were still considered unsafe. |

---

**Section Assessment**

1. Explain Anaximander's teachings which established belief in germ theory.
2. Explain the primary assumption about germs according to miasma theory.
3. What two common herbs used today were instrumental in the disinfections?
4. What significant discovery happened during a plague period in China which changed medical practices forever?

---

## 1.2 PART 2: HISTORY OF CARDIOVASCULAR MEDICINE

The following is an overview of some developments of cardiology in Western medicine over a 200-year period.

### 1.2.1 Huang di Nei jing

The Su Wen contains a number of passages concerning the heart and circulation, writing: "The blood current flows continuously in a circle and never stops." "All the blood is under the control of the heart." "The blood cannot but flow continuously like the current of a river, or the sun and moon in their orbits. It may be compared to a circle without beginning or end." These quotes helped to develop perspectives about the flow of qi and blood, yin and yang, and all syndrome types which identify diseases of the cardiovascular system.

## 1.2.2 Developments Through the Centuries

a. 1600s

- 1628 Exercitatio Anatomica de Motu Cordis et Sanguinis was a textbook which explained systems in medicine including the circulatory system and was written by William Harvey

b. 1700s

- 1785 Foxglove (digitalis) was first described by William Withering for use in strengthening and slowing heart rate and to treat dropsy (edema)
- 1799 Discovery of Nitrous Oxide (NO) as an anesthetic by Humphry Davy

c. 1800s

- 1819 Invention of the stethoscope by Professor René Laennec
- 1893 The first successful heart surgery was performed by Dr. Daniel Hale
- 1895 Discovery of the use of X-rays for medical imaging by Wilhelm Roentgen
- 1896 Development of the blood pressure cuff by Riva Rocci
- 1897 Invention of aspirin by Edward Stone

d. 1900s

- 1901 First report of cardiac occlusion with chest pain by Ludolf von Krehl
- 1903 Development of electrocardiograph for cardiac recording by Willem Einthoven
- 1905 Development of blood pressure with auscultation by Nikolai S. Korotkoff
- 1909 *Heart* was the first cardiovascular journal published in English
- 1910 Cholesterol in the aorta is first described by Adolf Windhaus
- 1913 Lipid theory of atherogenesis was described by Anitschkov and Chalatow
- 1924 The American Heart Association was started
- 1925 *American Heart Journal* was first started
- 1933 Group A β-hemolytic streptococci produces a fibrinolytic substance called streptococcal fibrinolysin was discovered by William Tillet and R.L. Garner
- 1934 The Belgian Cardiac Society was founded
- 1935 The Italian Cardiac Society was founded

- 1937  The French Cardiac Society was founded
- 1941  Plasma lysing factor necessary for streptococcal-mediated fibrinolysis was discovered by Haskell Milstone
- 1941  Cardiac catheterization was begun at Columbia University by Dr. Andre Cournand and Dr. Dickinson Richards
- 1943  Blue baby syndrome was cured based on hemorrhage and shock research by Dr. Alfred Blalock and his assistant Vivien Thomas, LLD
- 1945  The mechanism of streptococcal fibrinolysis was discovered by L. Royal Christensen
- 1948  The Swiss Cardiac Society was founded
- 1937  The British Cardiac Society started
- 1947  Invention of the Defibrillator by Claude Beck
- 1948  National Heart, Lung and Blood Institute was founded
- 1949  The American College of Cardiology was founded
- 1952  Atrioventricular block or advanced bradyarrythmia was first controlled by an external pacemaker by Paul Zoll
- 1953  The Heart–Lung Machine was invented by Dr. John Heysham Gibbon
- 1954  Echocardiography was developed by Edler and Hertz
- 1957  Benzothiadiazines were first developed
- 1958  Coronary arteriography was developed by Dr. Mason Sones
- 1958  Implanted cardiac pacemaker was developed by Elmqvist and Senning
- 1959  Aldosterone inhibitors were first developed
- 1961  The Framingham Heart Study first identified risk factors for healthy people with hypertension and hypercholesterolemia toward prevention
- 1962  Loop diuretics were first developed
- 1962  Invention of beta blocker medication by James W. Black
- 1962  Cardiopulmonary resuscitation (CPR) was developed to the procedures used today at John Hopkins University
- 1963  Invention of the artificial heart by Paul Winchell
- 1964  Coronary by-pass grafting was first developed by De Bakery et al
- 1967  The first human-to-human heart transplant was performed by Christiaan Barnard
- 1968  Coronary by-pass surgery procedure was changed to the saphenous vein to the internal mammary artery by George Green, Teruo Hirose, and Charles Bailey

- 1969  Invention of the balloon catheter by Thomas Fogarty
- 1971  Invention of magnetic resonance imaging (MRI) by Raymond Vahan Damadian
- 1974  A stapling technique for bronchial stump closure was developed by Dr. Richard N. Scott
- 1974  The Association of Black Cardiologists was founded by Dr. Richard Allen Williams
- 1974  The first peripheral balloon angioplasty was performed on a human by Andreas Gruentzig
- 1976  Evidence in support of statin use was developed by Akira Endo
- 1977  The Gruentzig procedure which is a technique of percutaneous transluminal coronary angioplasty (PTCA) was described in the *New England Journal of Medicine* and began the era of the development of interventional cardiology
- 1980  First human implantation of the automatic implantable defibrillator by Dr. Levi Watkins
- 1980  *European Heart Journal* was founded
- 1981  *International Journal of Cardiology* was founded
- 1988  Invention of the intravascular stent by Julio Palmaz
- 1990  Human Genome Project was begun
- 1998  Invention of stem cell therapy by James Thomson
e. Twenty-First Century
  - 2000  The handbook on the human genome was completed
  - 2001  The invention of telesurgery by Jacques Marescaux

## 1.3 PART 3: INTRODUCTION TO HEALTH COMMUNICATIONS FOR CHINESE MEDICINE PRACTITIONERS

### 1.3.1 The Basics

Health communication means teaching or assisting the community about important issues in medicine and personal health. The purpose is for shaping and modifying attitudes and behaviors. First, social scientists team with health care experts to form studies on the population to determine the most important issues. Next they study which groups of people need the most help. Many scientific theories are used to understand and predict behaviors in the population, and solutions are formed on how to convince people to adopt certain health practices. Public and private agencies will develop social marketing programs to get people to believe in and spend money on products and services which restore, improve, and maintain health.

Cardiovascular disease is a major cause of serious health problems and death in the United States and worldwide. Western medicine practitioners have dedicated decades to developing examination and diagnostic methods, clinical protocols, treatment procedures, and medicines. Despite these advances, the public is divided about the health outcomes. Some patients benefit while others who don't will require other options in addition to Western medicine care.

In the United States, Chinese medicine practitioners have a scope which allows the treatment of patients using acupuncture, herbs, injection therapy, tui na, and other modalities, and have a responsibility to use them to address health needs in the cardiovascular disease community. Now is the time to advance the profession by researching medical information, observing patients, and developing health literacy information which is necessary to shape attitudes and behaviors about Chinese medicine for cardiovascular care.

### 1.3.2 Health Literacy

Much of the work on health literacy has focused on problems and solutions in the health care context, such as communication between health care providers and patients, medication management, and informed consent. Health literacy, however, is just as important in public health.

Health literacy means being empowered with knowledge about prevention, present health state; worst and best case scenarios with diagnoses, treatment options, and prognosis or outcomes.

Chinese medicine practitioners should be prepared to:

- Increase education, skills, knowledge, and clinical experience in cardiology.
- Influence the community using what messages and information about Chinese medicine is available to assist patients. This includes but is not limited to what information on indications and contraindications of formulas and herbal ingredients is necessary to protect the public.

- Integrating Chinese medicine terminology into medical communication so that the medical and public communities can increase acceptance.
- Becoming more aware of the ways that cardiovascular disease affects people in different racial, cultural, and genetic groups.
- Manage, correspond, or team with Western medicine practitioners to assist with acute and chronic health conditions.
- Utilize the patient narrative and other medical information to assist in health care decision making.
- Develop skills in public health surveillance methods including incidence, prevalence, and risk.
- Avoid a general approach when explaining about Chinese medicine. Develop a greater understanding about cardiovascular for the public and gather feedback: What problems or symptoms are most difficult? What is most important in a treatment program? What would provide the most benefit for these patients?
- Develop specialized website, online and hard copy print material as resources. Partner with health care practitioners in other medical disciplines, as well as medical educators and journalists to spread Chinese medicine information which will increase patient literacy.

---

### Chapter Assessment

1. When was the Nei jing written?
2. Who were the scientists instrumental in the proof and development of germ theory?
3. What were Koch's postulates in germ theory?
4. What were the medicines created by Hua Tuo used for?
5. What is the name of the medical law created by Hua Tuo?
6. What two common herbs used today were instrumental in the disinfections?
7. What are the three statements from the Su Wen concerning the cardiovascular system?
8. What were the significant developments in the 1700s for cardiology?
9. What were the significant developments in the 1800s for cardiology?
10. What were the significant developments in the twentieth century for cardiology?

## NOTES

## CHAPTER 2

# Key Elements and Scientific Theories

## CHAPTER OBJECTIVES

After studying this chapter, you should be able to:

1. Explain the communication theories used to transmit health messages to the public.
2. Explain the public health research theories which explain how public behavior toward health is shaped.
3. Discuss how the Precede-Proceed Model has been used to develop programs which have helped reduce cardiovascular mortality.

## 2.1 PART 1: KEY ELEMENTS

In 1972, an article called "Doctor and Patient Communications" was written by Korsch and Negrete. This article described the importance of the relationship to improve health outcomes. Public health later became an interest to health care researchers who started programs. Scientific theories began to develop, epidemiological data began being collected, and research literature began getting published. Between 1975 and 1996, many conventions, organizations, and journals were established. The *Journal of Health Communications* was started in 1996. American universities began creating degree programs in public health.

Health Communications in Traditional Chinese Medicine Cardiology. DOI: http://dx.doi.org/10.1016/B978-0-12-800125-7.00002-1

| Communication theory | Definition |
|---|---|
| Organizational Communication | This theory includes the structure of communication within and between agencies and businesses. |
| Social Communication | This theory includes the method that organizations use to transmit health agendas. |
| Social Marketing | This theory carries the agenda to encourage public thinking, attitudes and behaviors. |
| Media Advocacy | This theory considers objectivity and neutrality in spreading public information. |
| Medical Journalism | This theory considers accurate medical information, which influences governments, organizations and physicians. |
| Edutainment | This theory is a method of transmitting information to the public, as a form of education and entertainment. |
| Interpersonal communication | This theory includes the many methods of communication between people, and organizational structures that depend upon one another. |

## 2.2 PART 2: PUBLIC HEALTH RESEARCH THEORIES

When public health became an interest to researchers during the 1950s to the 1980s, scientific theories began to develop. When public health became an interest to social scientists, they began research programs. These research programs formed the social theories and models which have further shaped society.

| Public Health Research Theories | Definition |
|---|---|
| Diffusion of Innovations Theory | This theory explains how information, opinion and behaviors are circulated amongst a population. |
| Health Belief Model | This theory explains how a belief influences patterns of behavior, whether the belief is right or wrong. |
| Social Cognitive Theory | This theory explains how having personal efficacy, self motivation and monitoring leads to success. |
| Theory of Reasoned Action | This theory explains how attitudes are predictors of behaviors. |
| Ideation | This theory explains how ideas are spread through peer approval and other factors that affect decision making. |
| Convergence Theory | This theory explains the recognition that all industrialized countries tend to be more alike, which diminishes the possibilities of conflicts between them. |
| Stages of Behavior Change Model | This theory explains the process that a hesitant target in a social system resorts to acceptance. |
| Communication for Persuasive Theory | This theory explains 12 steps for assimilating a particular behavior into a person's everyday life. |

---

## Section Assessment

1. What was the significance of the "Doctor and Patient Communications" article?
2. What is the common main idea of the scientific theories?
3. What is the common main idea of the public health research theories?

## 2.3 PART 3: MASS COMMUNICATIONS THEORIES

| Mass Communications Theory | Definition |
|---|---|
| Cultivation Theory | This theory explains the effect that TV viewing has especially when the images are similar to real life experiences. |

## 2.4 PART 4: MARKETING-BASED MODELS

| Marketing-based Models | Definition |
|---|---|
| Social Marketing | This theory explains how the principles that are used to sell products, can be used to influence attitudes and behaviors in other areas. |
| Integrated Marketing Communications | This theory explains how marketing is used to encode a brand's attention to create a demand for it. |

## 2.5 PART 5: MODELS FOR STRATEGIC BEHAVIOR AND SOCIAL CHANGE COMMUNICATIONS

Communication for behavioral impact explains how encouraging behavior change in a society can save lives (i.e., stopping the spread of communicable diseases, risk factors for cardiovascular disease, lifestyle modification).

The Precede-Proceed Model uses mnemonics to plan, design, implement, and evaluate health programs.

| P= | predisposing | P= | policy |
|---|---|---|---|
| R= | reinforcing and | R= | regulatory and |
| E= | enabling | O= | organizational |
| C= | constructs in | C= | constructs in |
| E= | educational | E= | educational and |
| D= | diagnosis and | E= | environmental |
| E= | evaluation | D= | development |

### Chapter Assessment

1. What was developing the most in public health during the 1970s?
2. What was the purpose of public health research programs?
3. What is the common main idea of the scientific theories?
4. What is the common main idea of the public health research theories?
5. What is the significance of mass communications theories?
6. The Precede-Proceed Model uses what memory system to plan, design, implement, and evaluate health programs?

**NOTES**

## CHAPTER 3

# Cardiovascular Disease Prevalence and Traditional Chinese Medicine

## CHAPTER OBJECTIVES

After studying this chapter, you should be able to:

1. Describe what governments around the world have approved to assist improving public health.
2. Discuss the recent prevalence and incidences of health behaviors of US adults and risk factors.
3. Describe the various cardiovascular diseases currently being treated using Chinese medicine in hospitals in China.

## 3.1 PART 1: HEART DISEASE

Governments around the world have health agencies such as the Centers for Disease Control and National Institutes of Health. Their missions are to provide public health and surveillance programs. Cardiovascular disease is treated using Western medicine methods of physical examination, diagnosis, and treatment. It is studied at institutions which only view Western medicine methods and practitioners only study and practice in Western medicine hospitals, clinics, and treatment facilities. Heart disease is a major chronic health condition in the United States. It is the leading cause of death among all racial and ethnic groups. Males and females display different symptoms of heart disease; however, there are common risk factors which increase the likelihood of developing symptoms.

Common risk factors are as follows:

- Uncontrolled hypertension
- Smoking habit
- Uncontrolled dangerous cholesterol levels

Heart disease is described in populations by its prevalence and incidence. Prevalence means how widely the disease occurs in a

Health Communications in Traditional Chinese Medicine Cardiology. DOI: http://dx.doi.org/10.1016/B978-0-12-800125-7.00003-3

population. Incidence means the frequency and magnitude in which disease occurs. Between 2000 and 2010, the prevalence of heart disease changed in the United States. Before that time more than 80% of adults between ages 40 and 60 had experienced one of the associated risk factors. Currently, the prevalence has decreased to around 50% equally for males and females. Decreases in males have occurred more prevalently in age groups between 40−59 and 60 + . For females, the decrease was found only among the 60 + age group. The reason for the disparities toward females may be that heart disease is still often misdiagnosed in women. The reasons disease incidences have decreased overall may be because of trending changes in attitudes and behaviors about health and more education about cardiovascular risk factors. According to health agencies in 2007, 38% of adults in the United States used complementary and alternative medicine, for any acute and chronic conditions. Perhaps within this figure is another reason for decline in cardiovascular disease prevalence.

## 3.2 PART 2: CHINESE MEDICINE AND CARDIOVASCULAR DISEASE TREATMENT

Chinese medicine is a system of medicine similar to and different from Western medicine. The similarities are that there are methods of studying basic medical sciences and theories, guidelines for prevention and algorithms for identifying disease signs and symptoms, diagnostic criteria, treatment principles, treatment methods, modalities, medicines, prognosis predictions, and research of medicines and methods. The differences include the terminology, perspective of viewing disease manifestation and state, examination and diagnostic methods, medicines and procedures in various treatments, and prediction of outcomes.

In China, Chinese medicine hospitals are nearly as advanced in protocols and medical procedures as Western medicine hospitals. Most cardiovascular departments of both types of hospitals utilize Chinese medicine with Western medicine treatments. It is common to find surgical departments for cardiovascular treatment in Chinese medicine hospitals, but the scope is often limited to interventional diagnosing and/or treatments. Currently, Chinese medicine treats cardiovascular diseases such as coronary artery disease, hypertension, heart failure, hyperlipoproteinemia, sinus rhythm disorders such as arrhythmia and sick sinus syndrome, infectious diseases which affect sections of the heart such as viral myocarditis, rheumatic fever, and endocarditis, vascular diseases such as thromboangiitis obliterans and Takayasu arteritis, as well as certain cardiopulmonary disorders. There are currently over 200 medicinal formulas in oral, injectable, and intravenous forms utilizing over 150 herbs. Patients can receive care either as an outpatient in the clinic or as an inpatient.

---

### Chapter Assessment

1. What was does disease prevalence mean?
2. What was does disease incidence mean?
3. What are the common risk factors for cardiovascular disease?
4. Discuss the Chinese medicine perspective of cardiovascular treatments?
5. List the cardiovascular diseases indicated most for treatment in Chinese medicine.

---

Log on at www.niambiwellness.com to access the companion course and quiz for Module 1.


## NOTES

# Interpersonal Communications

## CHAPTER 4

# Health Provider and Patient Types

## CHAPTER OBJECTIVES

After studying this chapter, you should be able to:

1. Discuss the significance of the 4 main provider types on patient care.
2. Explain the significance of the 4 main patient types and how their attitudes might affect the treatment process.
3. Discuss how to match the 4 main health provider types with certain patient characteristics.

## 4.1 PART 1: 4 MAIN HEALTH PROVIDER TYPES

### 4.1.1 Authoritative

This provider is a parental type who takes over your life who may not appreciate their authority challenged. They expect patients (or anyone around them) to be compliant.

Medical problems will most likely be compartmentalized into rigid, results-proven algorithms to direct diagnosis and treatment. Alternative options, suggestions, or referrals are often discouraged.

### 4.1.2 Institutional

This provider likely works in a large multidisciplinary clinic or hospital setting with a lot of employees. Providers may not remember you, because they rotate in day and on call night shifts where applicable. Patients may have trouble establishing a relationship with one provider. The reason is because that provider may not be available during future visits.

The provider on duty may not be interested in conversations about symptoms and complaints. They may also use few diagnostic tests and tend to make quick treatment decisions. These providers are only efficient with acute yet limited or chronic yet stable conditions.

Health Communications in Traditional Chinese Medicine Cardiology. DOI: http://dx.doi.org/10.1016/B978-0-12-800125-7.00004-5

### 4.1.3 Intellectual

Intellectual providers usually practice, teach, and research within their profession.

They are usually considered experts and have access to the latest research in their field.

These providers are most interested in patients with unusual or advanced diseases.

### 4.1.4 Supportive

This provider is intuitive to your emotional needs and uses touch as part of the warm gesture of caring. They spend a lot of time discussing details in consultations and physical examinations. They may work in a private practice and specialize in Oriental medicine. They may also practice in multidisciplinary settings, with flexibility and integration with other healing arts.

## 4.2 PART 2: 4 MAIN PATIENT TYPES

### 4.2.1 Skeptical

This type of patient likes to collect second opinions, yet discount all of them. This patient engages the provider in a stand-off battle involving willpower. The intellectual provider is best for this patient, because of the need to establish proof and an alibi. This kind of patient needs to know about research conclusions. They need to adhere to batteries of diagnostic testing to prove evidences. They also need definitive counseling about pros and cons of any treatments.

### 4.2.2 Intellectual

This kind of patient asks lots of questions and researches information on the Internet.

They usually bring medical records from all providers to piece together evidences to help solve their medical problem. The intellectual provider is best for this patient, because of the need to establish proof and an alibi. This kind of patient needs to know about research conclusions and multiple treatment options.

### 4.2.3 Flexible and Explorative

This kind of patient feels there is a lot of uncertainty in medicine and accepts minimal and complete care in treatment. A course of treatment is like embarking on the life journey. Personal history, life experiences, relationships, future goals, and preferences become part of the healing process. This kind of patient may feel drawn to try Oriental medicine. Both the institutional and supportive providers are best for this kind of patient.

Institutional providers can treat the acute or manageable chronic problems, while the patient explores more information. Supportive providers are often more open to the alternative or integrative therapies these patients request.

### 4.2.4 Passive

This kind of patient usually expects the provider to know what is best for them.

They usually comply with the provider's wishes even if the condition is not improving.

If they have questions, they usually rely on nursing staff or other medical staff for answers. The parental provider is the best for this kind of patient.

The provider tells the patient what to do, and they are expected to follow.

## 4.3 PART 3: 14 MAIN PATIENT CHARACTERISTICS

### 4.3.1 Manipulative

These difficult patients have learned certain behaviors that have resulted in getting selfish results. These patients don't consider themselves as manipulative and don't like being called this.

### 4.3.2 Demanding

These high-maintenance patients want more attention than their condition requires.

These patients may not be aware of their behavior.

### 4.3.3 Angry

May initially be disagreeable, but usually feels helpless and ignored. Managing this patient means controlling your anger during encounters and realizing their anger is short lived.

### 4.3.4 Noncompliant

These patients can be quite frustrating, because they usually don't follow your instructions or advice. This patient might be:

- Angry
- Manipulative
- Depressed

### 4.3.5 Depression

These patients are very easy to spot during interviews. Patients are not always open to providers pointing out this as a reason for their behavior. The problem is noticing any suicidal tendencies, which may be difficult to address in a single office visit.

### 4.3.6 Malingering

These patients may require regular treatment for real or imagined symptoms.

They may also be visiting because of some personal gain, such as personal injury payment, lawsuit.

### 4.3.7 Know-It-All

These patients have "just enough knowledge to be dangerous." Their limited knowledge on the subject makes them believe that they know a lot about it. They bring articles or quote favorite gurus to educate you, so you can be as knowledgeable as they are.

### 4.3.8 Geriatric Patients

These patients are chronically ill with several disabling diseases, often causing depression. These patients require a lot of empathy from their practitioners and caregivers for daily living.

### 4.3.9 Direct

These patients want to be in control and want to maintain this in the relationship.

They will tell you immediately that they agree or disagree with you and why.

### 4.3.10 Psychosomatic
Verifying causes of complaints and symptoms is difficult using diagnostic methods. Patients informed that their condition is psychosomatic can become disbelieving and upset.

### 4.3.11 Chronic Pain
These patients are considered a tragic challenge to providers. They may have miserable and crippling symptoms of a benign condition. It is difficult to determine between patients who really need or are addicts. Many providers worry about treating these kinds of patients. Providers fear legal consequences if anything goes wrong.

### 4.3.12 Terminal Patients
These patients have incurably fatal or malignant diseases. Western medicine providers tend to give up on these patients, feeling there is nothing more that can be done. Some of these patients choose to seek alternative treatments. Some providers may suggest treatments with unrealistic prognoses for these patients. This is the time when many of them might need long-term or hospice care.

### 4.3.13 Courageous
These patients show amazing characteristics when facing adverse circumstances.

They are admired for their ability to maintain strength, perspective, and acceptance.

### 4.3.14 Pleasant
This kind of patient is easy, as they seem compliant and manageable. These patients may or may not have malignant diseases. The concern is becoming too optimistic and accommodating, in light of adverse circumstances. Doing so may not be in the best interests of the patient.

## 4.4 PART 4: CONSTITUTIONAL THEORY

The constitutional theory is a method for the health practitioner to use patterns related to seasons of the year, environment, gender, body shape, physical features, emotions, intelligence, diet, work, recreation, spirituality, and pathological states. The practitioner can use this theory to help profile the patient and further personalize the examination, treatment principle, and treatments.

## NOTES

# CHAPTER 5

# Patient and Provider Communications

## CHAPTER OBJECTIVES

After studying this chapter, you should be able to:

1. Explain the periods in history when researchers began developing research programs to help physicians improve patient outcomes.
2. Discuss the types of social support given by health care providers to patients.
3. Define the support theories which explain health care and provider relationships.

## 5.1 PART 1: INTRODUCTION TO SOCIAL SUPPORT

Public health became an interest to social scientists between the 1950s and the 1980s.

Research programs investigating the physician and patient began to develop. These research programs formed the social theories and models which have further shaped society. In 1972, an article called "Doctor and Patient Communications" was written by Korsch and Negrete. Researchers soon developed strategies in communication to help improve patient outcomes. Physicians began to realize the importance of providing effective types of support.

| Types of support | Definition |
|---|---|
| Enacted support | This support type takes a dominate role over the patient's wellbeing, and is provided for the children and the critically ill. |
| Perceived support | This support type represents the perception of the level and quality of care from a physician. The patient expects empathy and sympathy from the practitioner. Recovers after illness, warms yang. |
| Empathizing | This support type involves understanding the patients' perspective. Sympathizing means demonstrating caring action. |

*Copyright © 2014 Anika Niambi Al-Shura. Published by Elsevier Inc. All rights reserved.*

Health Communications in Traditional Chinese Medicine Cardiology. DOI: http://dx.doi.org/10.1016/B978-0-12-800125-7.00005-7

From here, researchers began observing and investigating patients to form theories which would later predict health outcomes.

## 5.2 PART 2: SOCIAL SUPPORT THEORIES

Researchers explained support theories which could encourage physicians to have more patient empathy. Instructional programs began to develop based on the following theories.

### 5.2.1 Social Support Theory

This theory explains the importance of enacted support given by health practitioners. Effective enacted support assists patients in adapting to stress and solving problems.

### 5.2.2 Life Span Theory

This theory explains that a well-adjusted patient exhibits two traits that are linked to health state.

- Normal social relationships
- Personality

Social relationships develop throughout life and include parents and others.

The personality should be adaptive. Characteristics include low neuroticism and hostility.

It also exhibits high optimism, coping, and social skills. These skills are thought to be key for healthy doctor support perception. A healthy perception of physician enacted support assists in adapting to stress and solving problems.

### 5.2.3 Relational Regulation Theory

This theory explains the poor connection between patient's support perception and mental health.

- Patient's mental health is connected to their regular social interactions.
- Perceptions of physician enacted care are subjective to the health provider's behavior.

## NOTES

# CHAPTER 6

# Palliative Care

## CHAPTER OBJECTIVES

After studying this chapter, you should be able to:

1. Define the role of palliative care for certain patients.
2. Describe what palliative care is.
3. Explain marketing models and their intended effect on the community.

## 6.1 PART 1: INTRODUCTION

Palliative care is offered to terminal patients with chronic debilitating and malignant conditions which are beyond medical treatment. The purpose is for providing comfort and reduction of pain while the patient deteriorates toward impending death.

Palliative care teams include medical providers, mental health workers, and a chaplain.

## 6.2 PART 2: QUALITY OF CARE

Palliative care helps patients with chronic debilitating diseases that include kidney failure, congestive heart failure, and chronic obstructive pulmonary disease. Other malignant diseases include cancer, AIDS, Parkinson's.

## 6.3 PART 3: PROVIDER PARTNERING

The palliative care team plays an active and vital role during the end stages of a patient's life. The health care team can include medical professionals and nonmedical volunteers.

Health Communications in Traditional Chinese Medicine Cardiology. DOI: http://dx.doi.org/10.1016/B978-0-12-800125-7.00006-9

Service is not limited to patients and helps the caretaker as well. Caretaker burden is a very important yet poorly addressed factor in the care for the patient. Above all of their responsibilities and roles, the persistent illness and imminent death of a loved one are a very tough time for them. Caretakers experience high levels of stress and anxiety, which can lead to adverse health problems. Some common symptoms include digestive, cardiovascular, musculoskeletal, and sleep disorders.

Chinese medicine takes a holistic perspective on health care and should be readily available. In many cases, Chinese medicine will not be indicated for the patient; however, caretakers will always need options to help alleviate their symptoms of stress and grief. Certain methods in Chinese medicine can address the emotional and physical health. Acupuncture is already known to immediately address physical pain, anxiety, stress, and insomnia. Tui na addresses physical pain and stress. Herbal medicine offers an alternative to pharmaceutical drugs for insomnia, stress, anxiety, and improving mood. Chinese medicine treatments can be of service alongside others in hospice situations.

---

**Module Review Questions**

1. Discuss the significance of the 4 main provider types on patient care.
2. Explain the significance of the 4 main patient types and how their attitudes might affect the treatment process.
3. Discuss how to match the 4 main health provider types with certain characteristics. Explain the periods in history when researchers began developing research programs to help physicians improve patient outcomes.
4. Discuss the types of social support given by health care providers to patients.
5. Define the support theories which explain health care and provider relationships.
6. Define the role of palliative care for certain patients.
7. Describe what palliative care is.
8. Explain marketing models and their intended effect on the community.

---

Log on at www.niambiwellness.com to access the companion course and quiz for Module 2.

# SECTION *III*

# Caseload and Research

## CHAPTER 7

# Epidemiological Overview

## CHAPTER OBJECTIVES

After studying this chapter, you should be able to:

1. Explain how the Bradford Hill Criteria is used to determine causation.
2. Describe how research studies are used.
3. Define errors in validity and types of bias.

## 7.1 PART 1: INTRODUCTION

### 7.1.1 Correlation Does Not Mean Causation

Epidemiology is the study of the distribution and determinants of health and disease in populations. Clinical epidemiology includes the principles which are used to evaluate diagnoses and treatment methods in clinical practice. Biostatistics is a necessary section of epidemiology and applies relevant statistical theories and methods for medical sciences.

### 7.1.2 Bradford Hill Criteria

By 1965, Austin Bradford Hill had studied much about how one determines "cause." He then created nine criteria to determine this.

Health Communications in Traditional Chinese Medicine Cardiology. DOI: http://dx.doi.org/10.1016/B978-0-12-800125-7.00007-0

| Criterion | Definition |
|---|---|
| Strength | The larger the association, the more likely causal. |
| Consistency | The more consistent the findings over different sample factors, the more likely causal. |
| Specificity | The more specific an association, the larger the probability of cause. |
| Temporality | The effect must occur after the cause. |
| Biological gradient | The wider the exposure the more prominent the effect; inversely the wider the exposure the lower the incidence. |
| Plausibility | The more complete the knowledge of the mechanism, the more understanding of the plausibility of an effect. |
| Coherence | Some public findings that can be created in controlled settings. |
| Experiment | Evidence is proven through controlled study. |
| Analogy | Different studies may have similar effects. |

## 7.2 PART 2: RESEARCH STUDIES AND GOALS

| Research studies | Use |
|---|---|
| Observational Studies | Observing disease patterns and frequency to evaluate association between risk factors and outcomes. |
| Interventional Studies | Experiments and clinical trials which are conducted to explore the association between treatment interventions and outcomes. |

| Types of research | Definition | Goal |
|---|---|---|
| Basic Research | Explains the phenomenon being investigated. | To understand and explain a specific characteristic. |
| Applied Research | Applies a scientific explanation to real-world situations. | To understand problems and potential solutions. |
| Summative Research | Summates the cause and effect specifics of a study. | Summates the specific conditions under which an interventions was directly effective for a specific problem. |
| Formative Evaluation | Evaluates the effectiveness of programs developing in their settings. | Evaluates the effectiveness of a program within the setting of it's development. |
| Action Research | Studying specific problems which are active in a location. | Studying specific problems within a program, setting or organization. |

## 7.3 PART 3: KINDS OF STUDIES

| Types of studies | Use |
|---|---|
| Cohort Studies | Comparing patients with a risk factor or other exposure to others without them. |
| Case Control Studies | Comparing a case sample group with a disease with a case control sample group without the disease. Later a retrospective analyzes what potential exposures both groups encountered over time. |
| Case Series | A study on the experience of a patient, or small group of patients with a similar diagnosis; or comparing periods during which they are exposed to some factor which produce an illness with periods when they are unexposed. |

## 7.4 PART 4: VALIDITY AND BIAS

### 7.4.1 Validity

| Errors in validity | Definition |
|---|---|
| Random Error | Problems with the results of statistical calculations due to typos or how question or answer choices are interpreted by some participants. |
| Systematic Error | An error occurs when a malfunctioned instrument or device used in the study causes a different value in the study from the population. |

### 7.4.2 Bias

| Types of biases | Definition |
|---|---|
| Selection Bias | Excluding information because it affects the desired outcome. |
| Information Bias | An error in measurements where similar results cannot be duplicated. |
| Confounding | A hidden or unmentioned factor which explains why two variables have a positive correlation. |

## NOTES

## CHAPTER 8

# Research Methods for the Private Clinic

## CHAPTER OBJECTIVES

After studying this chapter, you should be able to:

1. Explain the development of Chinese medicine in the United States.
2. Define qualitative and quantitative studies.
3. Discuss the two methods health practitioners can use to develop a research program in their own practice and improve their own skills.

## 8.1 PART 1: INTEGRATIVE ORIENTAL MEDICINE PRACTICE

Traditional Chinese Medicine (TCM) is a term used to describe theories, practices, and treatments in Chinese medicine. Oriental medicine and Asian medicine are the terms used for the Traditional medical systems practiced throughout Asia. These systems share and also have their own unique acupuncture theories, practices, herbal pharmacopoeia, and methods for their uses in treatment.

### 8.1.1 Development of Integrative Medicine

Research and development of the integration of TCM with Western medicine started in China in 1978. The *China Journal of Integration of TCM and Modern Medicine* started in China in 1981. Researchers from Traditional Chinese medical universities throughout China have published literature in mainland journals and those of other countries. There are now more than 30 medical specialties and subspecialties in China that practice integrative medicine.

Chinese medicine is practiced and researched by many prestigious private and public institutions in the United States and worldwide. There are now over 50 different recognized international journals for Oriental medicine and integrative medicine with favorable impact factors. Hundreds of research literature with data that supports efficacy

Health Communications in Traditional Chinese Medicine Cardiology. DOI: http://dx.doi.org/10.1016/B978-0-12-800125-7.00008-2

and credibility are published every year. As a result, TCM has become an important system of medicine, especially in the United States. Institutions are private and regionally accredited medical universities, which are held to high standards of national and governmental agencies. Practitioners must complete a rigorous four-year masters degree program and pass national board exams before applying for a license to practice. Acupuncture and Oriental medicine are both credentialed in all states, yet some have wider practice scopes, clinical titles, and public awareness than others. Chinese medicine practitioners have postgraduate options to study further for their PhD in Oriental medicine, specializing in many areas such as sports medicine, gynecology, and oncology. Practice options range from private practices to collaborations with Western medicine practitioners, forming integrative referrals or group practices.

## 8.2 PART 2: RESEARCH METHODS

### 8.2.1 Method 1: Recruiting New Patients to Study

Which patients with specific diseases would you like to focus more on in your practice?

A good way to really connect with their needs is to conduct a small private study in your practice. This study may or may not be developed for publishing, but the aim of the study must be for personal educational purposes.

- Decide on a prevalent disease or personal characteristic you are interested in studying.
- Design a *qualitative* or *quantitative* study.

A *qualitative* study is also ethnographic. It allows the researcher to gather facts from a few selected participants about their behaviors. This helps researchers understand trends.

A *quantitative* study takes a small sample from a larger random population. It seeks to understand incidences or prevalence found in the greater population. One benefit is that patients can anonymously give their direct and honest views without feeling judged.

In your study, the patients could collectively be called a population or a sample from a population.

You should first prepare for your study by drafting a document based on the scientific process. Next tick the sections you cover as you learn about your subject and exploring related research articles. Next decide on an ethical way to select patients. One idea is to select a list from your own files. Another idea is to attract patients by volunteering in organizations devoted to the related medical problem. Have a minimum and maximum number of patients whom you want in the study.

| Scientific process | Definition |
| --- | --- |
| • Question | Queries what problem to solve. |
| • Hypothesis | Predicts the outcome of the study. |
| • Prediction | Gives further analysis of factors in the prediction. |
| • Test | Details the steps and procedures in the experiment. |
| • Analysis | Discusses the observations, outcome and conclusion of the study. |

Once you have the study population, you can provide surveys that collect data to help you understand patient opinions and needs. You can also screen, select, and treat the patients according to a published research study. You can treat them on a fixed day of the week for a specific period of time. Later you can arrange a data analysis using mining software. You can interpret the results yourself or hire a statistician/scientist. The results can focus your practice to serve certain patients more effectively. For example according to your data, you might decide to advertise according to certain trends found among your patients. You can also use deeper insight about a condition to convince people to become your patients.

## 8.2.2 Method 2: Studying Existing Caseload Files

Practitioners with a sufficient caseload of files can mine within them for valuable information. Depending on the size of your load you can either do the analyses yourself or hire a professional to mine your data. You can use simple math, accounting software, patient management software, and spreadsheets. For a more detailed investigation, you may need to use statistical analysis and data mining. Sometimes hiring a data mining professional may be relatively expensive, yet can save time and provide a deeper analysis. For example, you may find out how effective your prescriptions are across certain test groups.

This kind of information can assist in a self-assessment program and prepare for a more a targeted health campaign.

---

### Module Review Questions

1. Explain how the Bradford Hill Criteria is used to determine causation.
2. Describe how research studies are used.
3. Define errors in validity and types of bias.
4. Explain the development of Chinese medicine in the United States.
5. Define qualitative and quantitative studies.
6. Discuss the two methods health practitioners can use to develop a research program in their own practice and improve their own skills.

---

**Log on at www.niambiwellness.com to access the companion course and quiz for Module 3**.

This also concludes the Health Communications 1 course. It is strongly suggested that you log on to the courses at the companion websites to review the course modules. Next, submit course documents and complete the final exam.

Upon passing the exam, you will receive completion certificates which include your name and practice license number, along with the specific number of credit hours awarded for this course. Electronic transmission of CEU and PDA credits will be sent to NCCAOM and your state medical board.